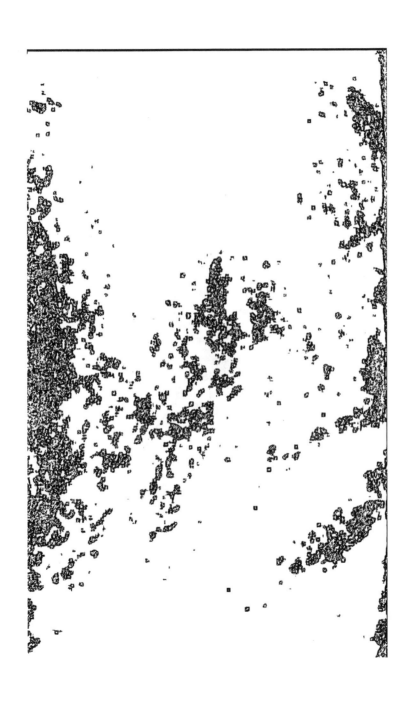

BROOM-CORN AND BROOMS.

A TREATISE ON

Raising Broom-Corn

AND MAKING BROOMS,

ON A SMALL OR LARGE SCALE.

Written and Compiled by the Editors of the American Agriculturist.

· · ·

ILLUSTRATED.

New-York:

ORANGE JUDD COMPANY,

No. 245 Broadway.

GARDENING FOR PROFIT:

A GUIDE TO THE SUCCESSFUL CULTIVATION OF THE

MARKET AND FAMILY GARDEN.

BY PETER HENDERSON.

Forcing Pits or Greenhouses.
Seeds and Seed Raising.
How, When, and Where to Sow Seeds.
Transplanting Insects.
Packing of Vegetables for Shipping.
Preservation of Vegetables in Winter.
Vegetables, their Varieties and Cultivation.

In the last chapter, the most valuable kinds are described, and the culture proper to each is given in detail.

Sent post-paid, price $1.50.

ORANGE JUDD COMPANY, 245 Broadway, New-York.

BROOM-CORN AND BROOMS.

A TREATISE ON

RAISING BROOM-CORN

AND

MAKING BROOMS,

ON A SMALL OR LARGE SCALE.

WRITTEN AND COMPILED BY THE EDITORS OF THE AMERICAN AGRICULTURIST.

ILLUSTRATED.

NEW YORK:
ORANGE JUDD COMPANY,
245 BROADWAY.
1876

CONTENTS.

———•◇•———

(3)

4　　　　　　　　　　CONTENTS.

INTRODUCTION.

Considering the importance of the Broom-corn crop, it is surprising how little is said about it in works on general agriculture. The literature of the subject is mainly confined to articles in the various journals, and the directions published by sellers of seeds and implements. Some of the latter give very meager instructions, showing a singular want of knowledge of the present methods of culture, while others are clever and useful treatises, though recommending implements and machines adapted only to the prairie soils of the Western States. In view of the demand for information upon the cultivation of this crop, we at first proposed to gather the various articles that have appeared from time to time in the *American Agriculturist*, and publish them as a pamphlet. The articles are by different editors and contributors familiar with the crop in various localities, from Maine to New York and Pennsylvania, and westward to Ohio and Illinois; but it was found that to reproduce these as they originally appeared, would involve a great deal of useless repetition, and one who wished information as to a par-

(5)

ticular point, would be obliged to refer to several different places. Instead of giving the articles as they at first appeared, we have consolidated the information given in all, as to the different operations, under the separate heads of "planting," "cultivating," "harvesting," and the like, which makes it much more convenient for reference.

Besides the articles referred to, we have embodied recent information obtained from correspondents, from dealers, broom-makers, and from the few publications that treat upon the subject. While it is in part founded upon the experience of those editors who have cultivated the crop, they only claim that it is a compilation from various available sources of information. To those who have kindly responded to our inquiries we return our thanks. THE EDITORS.

BROOM-CORN AND BROOMS.

BROOM-CORN AND ITS VARIETIES.

Formerly the Indian Millet, also known as Doura Corn, and by other names, the Chinese Sugar-cane and the Broom-corn, were regarded as distinct species of *Sorghum*, but botanists now look upon them as mere varieties of one species, *Sorghum vulgare*. The Millet has been grown in Africa, the East Indies, China, and other warm countries for centuries, and like other grains which have been so long in cultivation, is not known in the wild state. The different varieties are much unlike in external appearance and uses, but no more so than we find in some other cultivated plants; indeed there is nearly as much difference in the varieties of Indian corn; the sweet and popping kinds of which are unlike the tall-growing "horse-tooth" corn of the Southern States. When a plant is grown for a particular purpose, the cultivator endeavors to keep it improving in the direction most useful to him, by saving seeds from the plants best developed for his purpose. In Africa, Sorghum furnishes a large part of the food of the natives, and it is cultivated for its grain in various other countries, the sub-varieties of the grain-producing forms being numerous. In China and some other countries, where rice is abundant, this is less valued as grain, but has been cultivated for centuries for its sweet sap. The young stalk of the

(7)

grain-bearing kinds has a sweet juice, but this goes to form the grain, and soon disappears; by long cultivation with a view only to the quality of the juice, the seeds are less abundant and contain less starch, and while in the grain-bearing plants the head is so full and heavy as to bend over the top of the stalk and hang down, in the sugar-cane varieties the seed cluster is slender and usually erect; there are also sub-varieties of this. In the Broom-corn varieties, neither the grain or the quality' of the sap are the objects sought for, but a special and unusual development of the stems of the flower or seed-cluster; it makes no difference if it bear few or no seed provided these stems are long and fine. We cannot learn that *Sorghum vulgare* has been cultivated as a broom-making material in any other country than the United States until recently. It is said to be cultivated now for this purpose in Italy, France, and Germany.

DESCRIPTION OF THE PLANT.

Broom-corn is an annual grass with a general resemblance to Indian corn, but has narrower leaves, and instead of having its male and female flowers in separate places like the corn, with its tassels and ears, both kinds of flowers are in the cluster at the top. The flowers are of two kinds; one is perfect, *i. e.*, with both stamens and pistil, and seated directly upon the branch; each perfect flower is accompanied by an imperfect one, which is raised up on a little stalk; this consists either of empty husks, or bears stamens only, but can produce no seed, and falls early. The seed, or rather the husk which encloses the seed, is flattened-egg-shape, shining, with very fine hairs scattered over it. The stems of the panicle or flower-cluster are the valuable portion, all else being incidental; the main branches should be as uniform as possible in size, elastic and tough, long, and of a good color. Soil and cultivation, especially thick or thin planting, have much to do with the character of the product, the same kind of seed, with different soil and culture, producing brush greatly unlike in appearance and value. Besides these accidental differences, there are several varieties, in which by long cultivation and careful selection of seed, certain desirable qualities have become fixed.

THE VARIETIES OF BROOM-CORN.

Broom-corn culture was at the beginning of the present

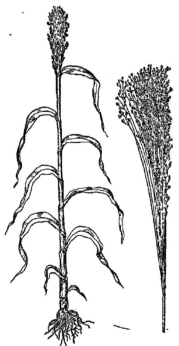

century confined mostly to New England and eastern New York, and the varieties known to the early cultivators have been superseded by other and improved sorts; the "Pine-tree," "North River," "Shirley," and other kinds are not recognized by our present growers.

Dwarf.—One of the most marked of all the varieties is the "Dwarf" (fig. 1), introduced about 20 years ago; the plant grows only about three or four feet high, nearly one-half of this being brush, of which it gives a large yield, but is more difficult to harvest than any other. The panicle, or brush, is partly enclosed by the sheath of the

Fig. 1.—DWARF BROOM-CORN.

upper leaf, while in the tall varieties it is usually raised above the last leaf on a longer or shorter stalk. At present it is only cultivated in limited quantities for making small brooms, whisks, and brushes for dusting clothes. The engraving of this variety is from the catalogue of R. H. Allen & Co., N. Y.

The Early Mohawk, from which the "Shaker" and "Early York" are not essentially different, is an old and still a successful variety, and though the brush turns red, it is very light, and on that account esteemed by some manufacturers of brooms.

The Tennessee, or Missouri Evergreen, are by some growers regarded as the same, while others think the Missouri rather the coarser. It is a tall growing variety, on good soil reaching 15 feet high, and at the great Broom-corn localities of the West is more esteemed than any other for producing a large yield of long fine brush, which seldom turns red if properly harvested. It is found that this, like other varieties, adapts itself to the locality, and when cultivated for a few years in succession, often gives more satisfactory crops than when the seed is brought from a distance.

"*California Golden*" is a variety offered for the first time this spring ; it is said to be in every respect superior to every other variety, claims which can only be substantiated by general culture.

INTRODUCTION AND EXTENT OF CULTURE.

Every one who writes on Broom-corn repeats the story of its introduction by Doctor Franklin, and this treatise would not be complete without it. It is said that Franklin, happening to see an imported broom, (some say whisk of corn), in the hands of a Philadelphia lady, had the curiosity to examine it, and finding a single seed, (others have it a few seeds), he picked it off and planted it ; this was the beginning of Broom-corn culture in this country. The story is used to illustrate the importance of cultivating the powers of observation, and the fact that small beginnings may often lead to great results.   As we do not know of any country, except the United States, that exports brooms, it would be very interesting to know where that broom came from, as it might throw some light upon the obscure history of this useful plant.

Like cork, Broom-corn is one of those natural products that are so perfectly adapted for the uses to which they are put, that no substitute has been, or is likely to be, found for it.   In toughness, elasticity, sufficient, but not too great rigidity, lightness, and ease with which it is manufactured, it excels all other materials used for brooms.   Besides these qualities, it is so easily procured as to be within the reach of every one, and can be grown over such a wide extent of country, that its culture, at least, whatever may happen to its commerce, can never be a monopoly.   However improbable the story of Franklin may be, its introduction is in keeping with the various benefits for which we are indebted to that eminently utilitarian philosopher.

Aside from the manufacture of brooms of various sizes and qualities, the uses of Broom-corn are but few.   It has one use which not many suspect, and this humble product often helps to complete the toilet of the most fashionable belle, and is even carried by the bride to the altar.   In making choice flowers into costly bouquets, they cannot be cut with stems long enough to allow them to be made up, and even if they could be, the natural stems are not so manageable as artificial ones ; accordingly each flower is first "stemmed," by attaching it by means of a fine wire to a splint of Broom-corn, which is the best material for the purpose.

The Shaker community at Watervliet, N. Y., are said

to have first made brooms for sale in 1798, though the plant was cultivated for home use some years previous to that time; Shakers and others in New England, especially in the valley of the Connecticut, soon engaged in the business to what then seemed a large extent.

As the demand increased, the cultivation extended westward, and for a long time the valley of the Mohawk was the great center of its cultivation, and there is still a large production of Broom-corn, as well as manufacture of brooms in that locality.

Some years ago Ohio produced large crops, and the southern and central portions of the State took the lead in this business.

At present the center of the culture is in Illinois, especially along the line of the Illinois Central Railroad, where not only is the aggregate area immense, but individual growers engage largely in the business, and it is not unusual to find from 300 to 700 acres in this crop belonging to one man. The methods of culture followed by these large producers are given in another chapter. Missouri produces a great deal of Broom-corn, but we have no accurate statistics for this or any other State, as strangely enough this important crop appears to have been quite ignored in taking the U. S. census for 1870. Some statistics, showing the extent of the manufacture of the product into brooms, are given at the end of the chapter on Broom Making.

It is probable that the returns for this crop in Kansas will be very large at the close of the present season, as the farmers of that State appear to be undertaking its culture very extensively. We learn from a friend in the Broom-corn district of Illinois, that the demand from Kansas for seed has been so great that the price has advanced to a point that may affect the breadth to be sown in Illinois this spring.

California has a climate and soils which seem to suit

almost all crops, and Broom-corn is found to succeed admirably there.

SECONDARY PRODUCTS.

SEED.—Brush of the best quality being the object of the cultivator, all other parts of the plant are subservient to this. To obtain the finest brush, it must be harvested when the plant is in flower, or at most when the seed is but slightly developed. Those who follow the instructions of one writer to harvest when the seed begins to ripen, may get a good crop of seed, but very poor brush. Formerly purchasers were not so exacting as to the quality of the brush, and the value of the seed was taken into account as a part of the returns for the crop, but at present one who wishes to produce the finest article allows only enough of the crop to mature to furnish seed for planting, as the diminished value of the brush is not compensated for by the value of the seed. Good, plump seed weighs 50 lbs. to the bushel, but the majority does not exceed 40 lbs. It contains a great deal of nutriment, and is by some regarded as equal in value with oats to feed to sheep ; ground, either alone or with Indian corn, it is an excellent food for pigs and for milch cows ; chickens eat it for a while as a variety, but are not fond of it long at a time. It will not, however, pay to raise seed for either of these purposes.

The seed from early cut brush hardly deserves the name, as it consists either of mere hulls with no kernel within them, or at most, hulls containing a seed from one-fourth to one-third developed The value of this will depend upon the degree of maturity. That from brush harvested very early and green, is worth no more as food than whatever nutriment may be contained in the hulls.

This seed, or rather hulls, for it is but little else, as it comes from the stripping machines, has been fed, more or less, to cattle and swine, but as it soon heats and spoils, the greater part of it has gone to the manure pile. Regarding this seed as too valuable to be allowed to go to waste, Prof. M. Miles, of the Illinois Industrial University, last fall made some experiments in preserving it. He stored it in pits, just as turnips or other roots are stored, putting on a layer of straw, and covering this with some 8 to 12 inches of earth. Pits put up in September were opened the following March, and were found in satisfactory condition ; where the covering was only eight inches thick, the outer portion was dry and molded, forming a compact crust a few inches thick, but the interior was fresh and bright, while a covering of twelve inches of earth preserved it better. A sample of this was sent us, and was found to be perfectly sweet, with much the odor of brewers' grains. What may be the feeding value of these immature seeds has yet to be determined, but there would appear to be no difficulty in keeping them perfectly well, should it be desirable.

THE FODDER.—In the large broom-corn fields of the west, the cattle are turned in after the harvest, and they are allowed to feed upon the leaves ; both these and the stalks are more nutritious when the crop is harvested quite green. No accurate comparisons have been made, but there is a general impression that the fodder of Broom-corn is about half the value of that from Indian corn.

THE STALKS.—Before flowering, the juice of the stalks contains a considerable amount of sugar, but this soon disappears, and the stalks after the harvest are quite dry and pithy. Where the dwarf varieties are cultivated, the stalks are plowed under, which gets them out of the way of the next crop and enriches the soil. This is not practicable with the tall kinds, which are burned to get rid of them, and their ashes are useful to the soil.

Where small lots are grown the stalks are sometimes taken to the barnyard to increase the amount of manure.

CULTIVATION.

THE LAND.—It is often said that any land that will produce a good crop of Indian corn, will answer for Broom-corn. The truth of this depends upon what is regarded as a "good" crop of Indian corn. There are many lands in the older States which might give the farmer a fairly remunerative crop of Indian corn, which would yield a very poor one of Broom-corn. In the Eastern States the quick and fertile, sandy, or even gravelly loams, such as are found in the river-bottoms, are most suitable. In the Western States the general fertility of the rich prairie soils renders a selection less difficult, and there are wide districts suited to this crop. The land should be as free as possible from weeds ; the Broom-corn while quite young is so small and delicate a plant that it is poorly fitted for a struggle for existence with weeds, and it is generally conceded that on land that is very weedy, almost any other crop will be likely to pay better than this.

ROTATION.—It is often found profitable to continue the crop on the same land, year after year. Where the weeds have been once subdued, the work of cultivation becomes less each year, and the land is so thoroughly shaded by the crop that new weeds have little chance to get a foot-hold. On rich prairie soils the land is kept in the same crop year after year. But most farmers who would grow only a moderate quantity, will make Broom-corn a part of their regular system of rotation, and turn under a clover or timothy sod to afford an excellent seed-bed for the crop.

Some advocate turning under the sod in June and keeping the surface free from weeds and mellow, by the

use of the harrow and cultivator, all the season. If done
for the benefit of future crops, this fallowing may be desirable, but one who followed this advice, expecting to
get his returns from the Broom-corn crop, would be
likely to be disappointed, as he has to compete with
those who can raise it at much less expense.

When a sod is turned over for the crop it is turned
deep, and the surface just before planting is pulverized
with a cultivator, or may have a shallow plowing with a
plow that can be guaged to the proper depth, and afterwards harrowed. This last operation should be done im-
mediately before planting, as it is all important that the
seed come in contact with moist, recently stirred soil.

MANURE.—Some cultivators upon prairie soils assert
that manure of all kinds is a great deal worse than useless, but elsewhere fertilizers are found to be essential.
The novice must be governed by what is required by other
crops, as Broom-corn demands quite as good treatment
as Indian corn, or other farm crops. Well decomposed
barn-yard, hog-pen, or sheep manure, made very fine, or
guano and other ammoniacal manures are used with advantage upon lands needing them. They are usually applied
in the drill, in order to forward the growth of the
young plant as rapidly as possible, with a view to get it
the sooner in advance of the weeds. Among special ap-
plications none have been so highly commended as plaster, which is claimed to have produced positively beneficial results when applied with the seed at the rate of
250 lbs. to the acre, and the application repeated at the
first working after the plants are well up. Ashes at the
rate of 12 or 15 bushels to the acre have been used with
great advantage, both alone and with plaster. Lime
upon soils where it is useful for other crops, may be applied with advantage to Broom-corn. Indeed no general
rule can be given to meet all localities, from those in
which the fertility of the soil appears to be inexhaustible,

to those on which, if nothing is put, nothing will be harvested. Whoever undertakes the cultivation of Broom-corn on an exhausted soil will make a great mistake if he follows directions given for localities where no field crops are manured, and expects that this demands any less from the soil than the other crops he has been accustomed to raise.

HILLS OR DRILLS.—As with Indian corn, some still adhere to the old method of cultivating in hills, so some still follow the same in growing Broom-corn, but the large cultivators sow it altogether in rows; the only reason for hill culture is, that very weedy land can be worked both ways; but this is just the kind of land upon which Broom-corn should not be sown at all. Those who sow in hills mark out the land 3 × 3 or 3 × 4 feet, with a corn marker, and scatter a dozen or more seeds, some using a teaspoonful—about 40 or 50 seeds—where the marks cross one another; when the plants are up, they are thinned to leave 5 to 10 of the thriftiest in the hill.

In planting in rows, the distance apart is governed by the character of the soil, the variety of the plant, and the quality of brush desired. Thick planting gives a fine and tough brush, but if the plants are too close it will be too slender. The rows vary from 28 inches to 4 feet apart, 3'|₂ feet being the usual distance for the tallest varieties on good land. It is usual to run the rows north and south in order that the sun may reach the plants more uniformly. In planting by hand, the rows are marked out by a small plow turning a shallow furrow, and the seed dropped about 2 inches apart, in a continuous row, or 6 to 10 seeds are dropped at intervals of 15 to 18 inches; some drop several seeds at intervals of 3 feet, when it becomes the same as hill planting.

In large fields hand-planting is too slow and costly, and planters or seed-drills are used. Where the amount to be sown is not very large, one of the several garden-seed drills may be used; the land being first marked off with a corn-marker. Whether sown by hand or garden drill in a continuous row, the seed must be covered by the use of a light harrow drawn lengthwise of the row; once only will be sufficient, unless the soil is hard and lumpy, which it should not be; if necessary to go over with the harrow twice, it should be run in the same direction as at first.

The various one and two-horse corn-planters have Broom-corn attachments, and may be quickly changed from Indian corn to Broom-corn planters, and some may be arranged to drop the seed 2 or 3 inches apart in a continuous row, or drop several seeds in a place at desired intervals; 6 to 10 seeds at intervals of 15 to 18 inches is a common method, while other large growers prefer the seed at equal distances. The machine at the same time covers the seed. A good two-horse planter, planting two rows at a time, will finish up 20 acres in a day.

The regular covering of the seed is of great importance, and in order that the machine may do it well, the soil must be very fine and mellow. The planter may be set to cover to any desired depth, which should never be less than three-fourths of an inch, or more than an inch and a half.

QUANTITY OF SEED TO THE ACRE.—This will of course depend upon the method of planting, and is stated at all the way from 2 quarts to half a bushel to the acre. Much of the seed contains many imperfect and poorly ripened ones, and allowance must be made for these, and for the chance that some may be covered too deep. The estimate that a bushel of good sound seed put in by a good planter, will plant 15 acres, (which is not far from 2 quarts to the acre), is a safe one, though it does not accord with that of a New England writer some 20 years ago, who thought a peck enough for an acre, but that some sowed a bushel, to be sure that the land was well stocked !

TIME OF PLANTING.—Being of sub-tropical origin, Broom-corn seed should not be sown until the soil is so thoroughly warmed that it will germinate at once. If put in too early, there is not sufficient heat in the soil to cause the seed to start, and it will either rot altogether and must be resown, or the young growth will be so weak that the weeds will rapidly get ahead of it. The time for planting Indian corn is usually given as the proper one for Broom-corn, but it may with advantage be a little later than that. In the Northern States it is planted from the middle of May to the middle of June. Of course the precise time will be governed by a knowledge of the peculiarities of the locality, as the crop must be harvested before the early frosts.

CULTIVATING.—Success with the crop, other things being favorable, depends upon keeping it free of weeds while young. The young plants when they first show themselves, are very small, appearing much like grass, and though they soon become strong and vigorous, they are weak at first ; the seed being, when divested of its hull, quite small, the germ has only enough nutriment prepared for its early growth, to enable the young plant to get fairly above the ground, and it has at once to form roots and provide for its own subsistence. In its young state it is poorly fitted to struggle with weeds, and unless these are removed from the start, the crop will be a poor one. Hence not only thorough, but immediate cultivation is required.

In order that the cost of production may be as low as possible, the crop is worked almost entirely by horse implements. To get ahead, and keep ahead, of the weeds, cultivation must begin as soon as the plants are well up, some commencing as soon as enough are up to allow the rows to be seen, and others waiting until they are two or three inches high.

IMPLEMENTS.—The harrow is the first implement used; a common ∧-shaped harrow with the front tooth removed, is drawn astride the rows, or some use a common two-horse harrow, run in the direction of the rows, finding that no more plants are uprooted than are beneficial to the crop; in either case the rows are harrowed once up and once down. Some western growers use a harrow for the purpose, provided with handles; this we pre-

Fig. 2.—HARROW-TOOTHED CULTI-VATOR.

sume is much like the harrow-toothed cultivator (fig. 2), an implement much used in the market gardens around New York, and until within a few years peculiar to them.

After this, whatever implement experience has shown to be efficient in destroying weeds in other crops, may be used. Sometimes the crop is worked, even from the first, with a light plow, but a horse-hoe or cultivator of some kind is preferable. The first time of cultivating, these implements are so set as to throw the earth from the row. At the second working these are set to throw the earth to the rows. The number of times the crop must be cultivated will depend upon the condition of the land and upon the season, but it must be done often enough to keep the weeds down until the crop can be laid by. Hand-hoeing will usually be necessary to remove the weeds that the larger implements have missed, and it is always required when the crop is in hills. This to be less expensive should be done while the weeds are yet small.

Implements of particular manufacturers are advocated

as being especially suitable for cultivating the crop; as these are probably no better than many others, we do not enumerate them. The intelligent farmer who knows that the success of his crop depends on keeping down weeds, will accomplish this end by employing those methods and implements with which he is most familiar.

THINNING.—If the planting has been so regular in the drills that the stalks stand 2 to 3 inches apart, no thinning will be required, but if thicker than this, the surplus must be pulled out. If planted in hills, only five or six plants should remain, the others being removed at the first hoeing.

HARVESTING.

WHEN TO HARVEST.—The quality, and consequently the value, of the brush depends in a great measure upon the time at which it is harvested. A delay of a week may make a difference of one-half, or even more, in the price in the market; yet important as it is, there is no one point upon which those who have written on Broomcorn differ more than as to the proper time to harvest, it being stated at all the way from blossoming time up to that when the seeds are ripe.

The buyers demand brush of a light green color, and though yellow or reddish brush may be really as good, it brings so much less in price that every care is taken to secure the desired green color. Some varieties are superior to others in this respect, but even with these, much depends upon the time of cutting and manner of curing.

The most successful growers say that the cutting should commence as soon as the "blossoms" begin to fall. After the flower has been fertilized and the seed "set," the anthers, or male organs and male flowers, fall away, and this is called the dropping of the "blossom." At this time the seed has just begun to form, and is in a merely rudimentary condition, and the brush at this period is not only of the best color, but is heavier—a matter of importance in selling—and it is thought to be more durable. The manner of harvesting differs with the variety, and the treatment to procure a very green product different from that where color is not so much regarded.

HARVESTING THE DWARF CORN.—Were it not for the great difficulty of harvesting, the dwarf corn would be much more generally cultivated than at present, as it yields more largely than any other, and its brush is better suited to some kinds of brooms. As stated in describing this variety, the base of the panicle, or lower part of the brush, is closely surrounded by the sheath of the upper leaf, or the "boot," as it is commonly called, and only the upper portion protrudes ; in a rainy season the accumulation of water in the boot often greatly injures the quality of the brush, which becomes very gummy, and soon turns red. On account of this peculiar manner of growth, it is found cheaper to harvest it by pulling the brush out of the boot, with a sharp jerk, than to cut first and remove the boot afterwards ; but at best the harvesting is troublesome, and especially so after the brush has become rain-soaked.

LOPPING, BENDING, OR BREAKING.—The old way of harvesting, when the seeds were allowed to ripen, was to first lop the brush. In this operation a man goes through the rows and breaks down the tops, bending those of two rows towards each other ; the distance below the brush at which this is done varies from a foot to 18 inches. Though called "breaking," the stalk is not severed, but bent over at as sharp an angle as may be, without actually breaking it off. The object of this operation is primarily to keep the brush straight; as the increasing weight of the seed, if this were not done, would cause the brush to curve and become permanently crooked, but this cannot happen when it hangs perpendicularly downwards.

Another object in lopping is to accelerate the ripening of the brush. After bending down, the growth in length ceases, but the top still retains sufficient communication with the stalk to allow it to harden and mature. It is generally practised in northern localities, where the season is not long enough for the crop to escape frost without this treatment. One recent writer upon the cultivation of Broom-corn, advises bending down as soon as the head is fairly developed, and going through the field several times to do this, as the plants successively come into the proper condition. Lopping is not practised by the large growers in the Western States; indeed where the corn is 12 and 15 feet high, it would be a difficult and expensive task.

CROOKED BRUSH sometimes causes loss to the grower, and will occur in some seasons when the crop is harvested quite green. When cut as soon as the "blossom" falls, although the seed is but little developed, and the weight from the increase of the kernel is but slight, yet the weight of the envelopes to the seeds, or hulls, already considerable at first, increases as they grow and become firmer. In dry seasons the brush remains straight, the straw as it develops becoming sufficiently hard and firm to sustain its weight. On the other hand, in hot and moist weather the development of the head is very rapid, and the straw does not become firm with sufficient rapidity to enable it to keep erect, with its constantly increasing weight, for in this weather the seed-hulls also grow more rapidly, and the consequence is, the straw bends over and forms crooked brush. Even in dry seasons, thinly planted corn has a greater proportion of crooked brush than that planted closely, and one can only learn by experience what distance between the stalks will, on his particular soil, best prevent crooked brush, and give him the finest quality in all other respects.

TABLING.—This operation is preliminary to cutting, and brings the tops down within easy reach of the cutters, a point quite essential with the tall-growing varieties. It consists in breaking down the stalks of two rows

Fig. 3.—TABLING AND CUTTING BROOM-CORN.

towards one another, diagonally, so that the stalks of one row will cross those of the other, and thus form a sort of platform or "table," with the tops projecting about a foot on each side, as shown in figure 3. This should be at a hight most convenient for the cutters, which is usually about 30 inches from the ground, at which hight the bend is made. Each two rows in the field are tabled in this way, the intermediate spaces affording room for the cutters.

CUTTING.—The directions for cutting, especially with reference to the time, given by the different writers upon Broom-corn culture, are remarkably at variance, as are those for its treatment after it is cut. When the tops are lopped, the brush is cut as it hangs, cutting so as to leave 6 to 8 inches of stalk below the brush. If a portion of the leaf is cut with the brush, it must be taken off. When the cutter has a handful, he lays it between two rows, usually bringing the brush from two other rows to this, forming gavels ready to be loaded upon a wagon to take it to the drying shed. Another method where no drying shed is used, is to cut the brush from two rows, then cut up the stalks from these rows, and lay them on the ground crosswise of the rows to form a sort of foundation for the brush ; that which is cut from 8 or 10 rows being laid here to be treated as mentioned under curing.

With tabled corn, the brush, as it is cut, is laid upon the tables, and is removed to the drying sheds as soon after as may be. The length of stalk left upon the brush is not a matter of so much importance to those who make up their own brooms, as to those who sell the crop. It being sold by weight, the grower naturally wishes to include all the stalk that the buyer will accept. On the other hand, if the buts are unduly long, the purchaser will demand a reduction in price, which will more than offset the gain from extra weight. Custom has fixed upon 8 inches as the proper length.

Each cutter has his fancy as to the best kind of knife ; some use a rather large knife, like a butcher's knife, while the majority prefer a lighter one, like a shoe knife, with a round point. Some cutters think a knife that is not very sharp preferable to one with a keen edge, as they can better avoid cutting off the leaf.

PREPARING FOR MARKET.

CURING THE BRUSH.—Notwithstanding the fact that the price is governed by the color of the brush, and that exposure injures the color, some still cure their corn in the most careless manner, an account of which is given here only as a method to be avoided, and to show what poor guides are some of the published instructions. Under cutting is mentioned the plan of laying the brush on beds of stalks between the rows ; this is left in the sun for two or three days, then tied up into bundles and stacked in small round stacks, which are covered with stalks, laid on in such a manner that the top will shed rain, while the air can pass through below. It is left in this manner two or three weeks, to be *cured*, though with this treatment for all profitable sale, it is more likely to be effectually *killed*. The writer who recommends this is one who does not cut his corn until the seeds are nearly ripe.

Even for home use, the brush should be cured under cover, as exposure renders it brittle, and without that toughness and elasticity which we look for in a good broom. Those who raise only small quantities can easily dry it under some shed, or other out-building. All that is required is a roof to cover it, and a free circulation of air. Before describing the building for drying, we will consider

THE SCRAPING, OR REMOVAL OF THE SEEDS.—To clean a small quantity of brush for one's own use, several simple devices will answer. A wooden comb made by sawing teeth in a plank, will do the work. Small lots may be cleaned by using a long toothed curry comb. A sort of three-toothed hatchel was formerly in use, before the invention of cylinder scrapers; this is made of three strips of elastic wood, or of iron set upright, as in figure 4. These three teeth are fastened into a lower plank, and pass through a hole in a second inclined plank. By introducing wedges between the outer teeth and the sides of the hole, the teeth may be rendered more or less firm. The operator draws the brush, by small hand-

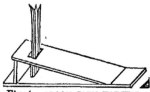

Fig. 4.—BROOM-CORN HATCHEL.

fuls, down through the teeth, which should be brought close enough together by the wedges to remove the seeds without breaking the straw.

Small crops have been threshed with a flail. The brush is laid upon the barn floor, two or three stalks deep, and a plank laid over the stalk portion to prevent crushing, the thresher standing upon the plank while swinging the flail.

Some remove the seed by the use of a threshing machine. The top or concave being removed, the brush, as much as can be held with both hands, is exposed to the action of the drum.

For large crops, special scraping machines are used, run by horse or other power. These are essentially one or two cylinders furnished with iron teeth, made to revolve very rapidly. Figure 5 represents a one-cylinder machine; with this the brush must be turned, in order to expose both sides; where there are two cylinders, revolving in opposite directions, the brush is held between them, and both sides are scraped. Two feeders can work at one machine. The brush as taken from the wagon is laid upon a long table, and one or two boys, according to the rapidity with which the feeder works, makes up the brush into convenient handfuls. When cleaned, the brush is thrown into a box behind the feeder, made like a saw-buck, with the sides and one end boarded up ; the ends of the stems coming against the back of the box, are kept straight. The brush is removed from the box to the drying house.

In this as in all other handling, it is very important to keep the brush straight and smooth.   It is very easy

Fig. 5 —SCRAPING MACHINE.

by careless handling to get it in a tangled condition, which materially decreases its value.

ASSORTING THE BRUSH is a matter of importance; as with most other products, when good and bad are mixed indiscriminately, the whole will sell for only the price of the bad.   Hence the poor and crooked brush should be separated, and kept distinct until it is baled for market. While the best growers agree in doing this, they do not agree as to when is the best time for doing it.   Some do the assorting when the brush is delivered at the scrapers, while others find it more advantageous to cull it before it is removed from the tables.   A man goes along in advance of the wagons, and places the poor and crooked brush by itself, and both the straight and crooked are stacked separately in the wagon and kept apart in all future operations.

THE DRYING OR CURING HOUSES.—It has been stated that for small crops, almost any shelter may serve, but large crops require ample accommodations, as the price will depend much upon the proper drying. On farms where a change has been made from tobacco to Broom-corn, the tobacco houses may be easily so arranged as to answer for this crop. Still, as a general thing, growers put up special drying houses. These are frame buildings with a tight shingled roof, the sides are covered with up-right boards a foot wide, with the joints battened, or covered with narrow strips; every fourth or sixth board is hung upon hinges, so that they may be opened to allow of a free circulation of air; provision must be made for fastening them by a button or otherwise, during a storm. For the sake of durability, the building should be painted, though this is often neglected. One grower estimates that 50 acres of Broom-corn require a building 20 × 40 feet and 16 feet high, with a shed upon one side 10 feet wide and 8 feet high.

RACKS FOR DRYING.—Racks are fitted up inside of the building upon which to place the brush to dry. These are built where timber is at hand, with poles for uprights, or light scantling 2 × 2 inches is used; where these are not the most suitable and cheapest, oak plank may be sawn into strips 1 × 3 inches. Whatever the material used for the uprights, they should be 12 feet long. Each pair of poles has narrow strips 4 feet long nailed to them, 6 inches apart, to form a sort of ladder; if good mason's lath can be had, these may be used; these are 3 feet 10 inches long, and if free from knots, will be strong enough. These racks are then set up on the floor of the house, 3 feet 10 inches apart; figure 6 shows the lower part of one of the racks. Other laths are laid across the strips, upon which the brush is to be placed to the thickness of about 2 inches.

CURING.—Such a mass of partly green vegetable matter as is thus brought together in a drying-house, will quickly heat in damp weather, hence the brush

Fig. 6.—RACK FOR DRYING.

should never exceed 2 or 3 inches in thickness upon the drying racks. Much will depend upon the state of dryness when it is put in, and upon the weather probabilities. The brush must be as dry as possible when put upon the racks, on which account cutting should not commence in the morning until the dew has dried off, otherwise the crop may be greatly damaged.

The time required to dry will of course vary with the weather ; it should be facilitated by attention to the ventilators, admitting all the air possible on dry and fine days, and shutting them during a storm. A look-out should be kept for sudden showers, and it should be the business of some one to attend to the shutting of the ventilating doors whenever there is danger of injury by rain.

DRYING AND HANDLING THE CROP ON A LARGE SCALE.

The form of the drying-house is of little importance, so that the requisite facilities for unloading the brush, piling it upon the racks, exposing it to currents of air,

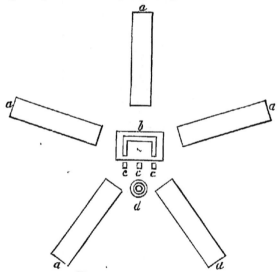

Fig. 7.—PLAN OF DRY-HOUSE.

and at the same time preserving it from rain and damp, and for removing it when cured, are provided. Probably one of the most convenient and practically useful drying arrangements in the country is that of Messrs. Johnson & Bogardus, near Champaign, Illinois, whose farm of six hundred acres, is all used for the cultivation of Broom-

corn of different kinds. The following description
of their buildings, and general management in harvesting, will show how admirable are the arrangements
for handling their immense crop. The dry houses,
of which there are five, are arranged around a central
building used for a sorting-house, in such a manner (see
fig. 7), that the wind has a free sweep in any direction,
and access can be had with wagons to each dry-house and
all around the sorting-house. The dry-houses are
shown at *a, a, a*, the sorting-house at *b*, the scrapers

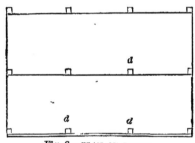

Fig. 8.—PLAN OF STALLS.

at *c*, and the horse-power—a 10-horse lever power—
which runs the scrapers for cleaning the brush, at *d*.
Each of these dry-houses consists of twelve sections or
stalls, as seen at figure 8, eight feet long, or seven feet
four inches in the clear between the posts. A house may
consist of any number of these sections needed to contain
the brush raised upon the farm, from one upwards. In
these five buildings there are 60 stalls (12 in each), which
are sufficient to hold the brush produced on the 600 acres.
The stalls are, however, filled three times ; first with the
early dwarf varieties, which are baled and out of the way
by the time the first of the later varieties are ready for
cutting ; these are dried and baled by the time the latest
come in. The season of harvesting is lengthened in this
manner, by planting succeeding varieties, as well as by

planting each variety in succession. A stall 8 feet wide,
24 feet long, and 16 feet high, is sufficient for 3 to 4 acres
if filled only once. The stalls are made by placing posts
4 inches square and 16 feet long, 8 feet apart from the
centers, as shown at *d, d,* in figure 8, and *d, d, d, d,*
in figure 9. These posts form a bent of the dry-house,
and each bent forms a stall. Laths or strips, one inch
thick, and two inches wide, are nailed to the posts 6 inches
apart from center to center, thus leaving spaces between

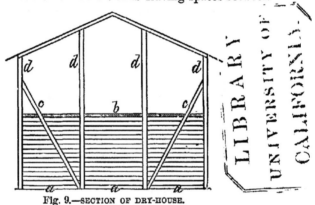

Fig. 9.—SECTION OF DRY-HOUSE.

them 4 inches wide. Movable laths are placed upon
these strips to hold the brush ; these are 8 feet long, thus
reaching across one bent. The laying of the brush is
begun at the middle of the stall, and as one tier of laths
is covered, another is placed and covered, the last brush
being put in from the outside of the building. The dry-
houses are open upon all sides to give free circulation of
air amongst the brush. Figure 9 shows the end eleva-
tion of the dry-house, *a, a,* being the strips which are
nailed to the posts *d, d ;* *b* is a floor over which is the
upper story used for storage ; *c, c,* are braces of 4 × 4 timber to strengthen the stalls. The houses are roofed with
boards. Figure 10 is a plan of the sorting-house. It is

surrounded upon three sides by a platform *a, a,* upon which the brush is unloaded from the wagons; within, also upon three sides, is the sorting-table *b, b,* at which the sorters, all Dutch women, work inside. The brush is sorted before it is cleaned, and is taken from the sorting-table to the cleaning-table *c,* in front of the machines

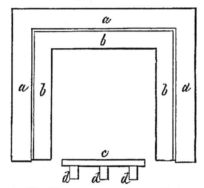

Fig. 10.—PLAN OF SORTING-HOUSE.

or scrapers, of which there are three, on the open side of the building, seen at *d, d, d.* The seed is collected from the scrapers into a pile, from which it is removed with carts.

BALING.—After the Broom-corn is thoroughly dry, the next step is to bale that which is to be sent to market. This operation should receive great care and close supervision, as the salableness and price of the crop much depend upon the appearance of the brush when it reaches the market. This may be said of every product of the farm, from strawberries up to apples and Broom-corn; the appearance sells it. One not familiar with the ways of markets, cannot understand why his neighbor's Broom-corn, grown on the same kind of land, in the same manner, and put into the drying sheds in precisely the same condition as his, brought enough more than his own to make all the difference between profit and no profit. Had he been in the city where the crops were sold, and seen the two lots on their arrival, he would have found that his neighbor's corn was delivered in compact, tight bales, with square, even ends, and that it was so closely packed that at the end of a long journey it was difficult to pull out a sample. On the other hand, his own bales were shaky, lop-sided, and instead of its being difficult to pull out a handful anywhere, it troubled those who handled it to keep it from falling out altogether. No matter how carefully and successfully every step in the production of the brush has been performed, the profit of the crop will depend, other things being equal, upon the proper baling.

THE PRESS.—Several presses are made for hay, cotton, etc., almost any of which can be adapted for baling

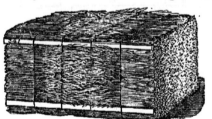

Fig. 11.—A BALE WITH LATHS.

Broom-corn, and some are constructed with special reference to this crop. They are made to work by horse-power or by hand. Where the crop is small, any one with mechanical skill can devise a press that will answer. The proper size of the bales is 3 feet 10 inches long, 24 inches wide, and 30 inches deep. The bale is bound by four or five wires, that known as "No. 9 fence wire," being the kind generally used. Some place a stout lath at each corner to protect the brush and to strengthen the bale, and some add a light wire which passes around the bale lengthwise, to hold the cross wires in place. Figure 11 shows a bale as commonly finished. This method of baling was more practised a few years ago than at present, it now being rare that bales reach the eastern markets with laths at the corners, or with end-wires. Four or five wires around the bale, in one direction, are sufficient to keep the whole in proper shape, provided the brush is properly lapped in the interior, and the buts placed evenly at the ends. If a bale thus made up is well pressed, it will reach its destination in good shape. Bales of the size above stated should, if properly packed, contain at least 300 lbs., and it is better for them to run a little over than to fall short of this weight. While bales of this size store better in the freight cars, and are the most desirable in the market, those as small as 150 lbs. are frequently sent, and sometimes they weigh as much as 450 lbs.

While the inferior, or crooked brush, should never be put in the same bale with the better quality, it is customary to select some of the best and straightest brush to give the outside of the bale a neat appearance.

It being of great importance to keep the ends of the bale square and smooth, the brush should be handed to the packer in small lots, the buts of which have been evened by striking them down upon a table or other smooth surface, and the one who places the brush in the box of the press, should take care to keep the buts up close against the ends of the box.

Bale the crooked brush by itself, and sell is as second quality. Recollect that a few crooked heads will injure the sale, and reduce the price of a whole bale otherwise of the best quality. A large amount of the best straight brush with the crooked, will not increase its market value, and the preliminary sorting, whether on the tables or at the scraper, should be entrusted to a faithful hand.

MARKETING.—The principal markets for Broom-corn are New York, Chicago, Philadelphia, and Baltimore. Some few commission merchants devote themselves exclusively to this article, while many general commission merchants deal in it largely. The commissions, cartage, storage, and other expenses, materially reduce the profits, and those growers are fortunate who can, as many do, make a contract directly with the manufacturers of brooms, even at a price less than the current quotations, as the saving of the various charges made upon consignments, more than compensates for the difference in price.

It is the misfortune attending the cultivation of special crops, such as Broom-corn, Hops, Oil of Peppermint, etc., that they are made the subject of speculation. It is true that this happens with the general crops, such as wheat, but not to the extent that it does with those in the cultivation of which but few persons are engaged. With these special crops it is easy for speculators to learn almost the exact number of acres under Broom-corn in the whole country, and the probable amount of the crop, and they sometimes manage to control very nearly the whole product. All that the farmer can do is to avoid entering into any combinations, or any agreement to hold his crop, but to sell whenever the price seems to him a fair and paying one.

COMMISSIONS AND CHARGES.—The following are about the usual charges made by commission merchants upon consignments of Broom-corn in the New York market:

Car-loads, sold without handling, a commission of..2½ per cent.
Broom-corn sold from the store, commission of.....5 " "
Guaranteeing sales on time........................2½ " "
Storage for each bale, per month...................7½ to 15 cents.
Cartage per bale (average).........................15 "
Labor in handling, per bale (average)..............20 "
Insurance varies from.............................1 to 2 per cent.

PRICES.—The prices in the New York market for the last five years have averaged as follows:

	Poor.	Good.
1871	3@10	8@15
1872	2@ 6	7@13
1873	2@ 6	8@14
1874	4@ 6	9@14
1875	6@ 9	9@15

The extremes for the first quality during this time being 7@15c., while those of the poorer grades were 2@10c.

PROFITS OF THE CROP.—In some years planters have made very large profits upon their Broom-corn; these unusual cases are widely published, while but little is said of the many growers who have found it only a fairly remunerative crop, and the losses from unfavorable weather or want of proper knowledge, never find their way into print.

We have before us the statements of several experienced growers in the Western States, where the character of the soil allows the crop to be produced at the minimum cost, these estimate that a ton of Broom-corn which is the average product of three acres, costs from $40 to $60. This at even the minimum price is a paying crop, and on a good market, the profits are still more satisfactory, but they are rarely extravagant.

In a year of large crops the careful grower has a great advantage, for as in all other products, the inferior qualities are first to decline, and they fall more in proportion than the better grades.

GENERAL CONCLUSIONS.—While those who have suitable land, and give the crop the attention it requires, find Broom-corn, one year with another, as profitable as any of the staple products, those who have read of the exceptional cases, and expect to accumulate sudden wealth by the culture, will be disappointed. Only those will find it profitable who have the best implements, buildings, and other appliances for carrying on the business year after year.

The great trouble with our farmers in undertaking any

special crop, is their fickleness. Easily elated by the accounts of success, or the profits that have been realized by a neighbor, they undertake a new culture without properly learning its requirements ; they abandon their regular crops, which they know all about, and put all their land into one about which they know nothing, and often invest all their capital in starting the new crop, staking all upon its success. Their first crop, upon which they have depended, may fail on account of unfavorable weather, or for want of proper management, or hundreds of others may have also gone into the same specialty, and the supply is increased so far beyond the demand that prices fall far below the cost of production, and the crop brings the grower in debt. The next year 19 in every 20 who have started in this new enterprise, will abandon it, but the 20th will try again, and ultimately succeed. The history of every special culture illustrates the above, and none more strikingly than that of hop culture several years ago in Wisconsin, where such a breadth of land was set in hops that it was impossible to procure a sufficient force to harvest the crop, and many prosperous farmers were ruined. As one who had never grown any other crop than wheat would make a sorry failure should he change to Indian corn without any experience or instruction, so it is with Broom-corn and other crops requiring special treatment in growing, and particularly in harvesting. The one who starts with a few acres, and when from these he has learned how his land is suited to the crop, and as he gains experience, gradually extends the area, taking it up as a part of his regular business, and making every investment in buildings and implements with a view to the future, will be likely to make Broom-corn pay. Those who without knowledge or experience abandon tried crops for this untried one, and invest in hundreds of acres instead of tens, will be very apt to find that there is "no money in it."

GROWING ON THE LARGE SCALE.

The following is from one of the most capable and experienced agriculturists in the country, who resides in the celebrated Broom-corn district of Illinois, of which Champaign is the center. His acquaintance with those who cultivate most extensively, his familiarity with the various plantations, and his general knowledge of agriculture, make his report of more value than that of any one grower. ·The very copious notes he has kindly furnished are here condensed.

The largest cultivators in this vicinity are Johnson & Bogardus, who have 600 acres in Broom-corn ; two other parties have 500 acres each ; two others 300 each, while those who have from 40 to 100 acres, are numerous. I am informed that south of here there are those who have still larger fields, but I am not able to give data. It is estimated that in this immediate vicinity there are 5,000 acres in Broom-corn. The growers generally own the land, though some hire it.

The Early Mohawk, and the Missouri or Tennessee Evergreen are the usual varieties, the Dwarf being but little grown. The seed is sown by means of two-horse drills in rows 3 to $3^1|_2$ feet apart, from 6 to 10 seeds being dropped every 15 to 18 inches. While some prefer 10 seeds, others plant only 6 in a place, and have the drills arranged accordingly.

In cultivating, the work is done all in one direction, there being no crossing of the rows. A cultivator is used which has adjustable scrapers so set as to throw the earth from the row, and in hand-hoeing the earth is worked from the row. Clean cultivation is essential, and whatever implements will keep the weeds down in other corn, are used for this. The amount of cultivation depends greatly upon the season, in *good weed years* it is more than in others. One man with a span of mules is expected to do all the cultivation of this kind, for 50 or 60 acres. Hand-hoeing follows the cultivating, and the amount of this that one man can do in a day depends much upon the condition of the land, and is all the way from half an acre to two acres.

As a general thing no manure is used, and when applied at all, only a very light dressing is given, as it is found that heavy manuring causes a stiff and rough brush. The same land is cultivated in Broom-corn year after year, there being some farms here upon which it has been grown for from 15 to 25 years successively without any diminution in quantity or deterioration in quality. Last summer I was through a field which had been in Broom-corn for the past 15 years, without intermission, yet many of the stalks were 15 feet high.

In harvesting, the stalks of every two rows are broken down into "tables," they are so broken that the stalks of one row will cross diagonally those of the other, and thus form a kind of table upon which to lay the brush when it is cut. In the northern part of the State, where they grow the smaller varieties, the tops are lopped and cut.

The seed is stripped off by machines, and the brush is cured in sheds with movable racks. The cutting is done when the corn is in "blossom," or at the latest, when the seed is in the "milk," in order that the brush, when cured, may be green. Some think that the brush is bleached ; this may sometimes be done with brush which has turned red, but the growers here all endeavor to secure a green color by early cutting and careful drying.

Here the seed mostly goes to the manure heap. Some feed it to cattle and hogs, but as a general thing it is practically wasted. It is much the same with the trash, or what is left after the harvest. It is usually burned to get it out of the way ; this of course fertilizes the soil to some extent, though that is not the primary object in so

doing. Your questions as to cost and profits are not so
easy to answer, as those who are very free in telling every-
thing else in relation to the crop, do not care to let their
neighbors see their day-book and ledger. The average
returns are probably from 500 to 600 lbs. of brush to the
acre, though we often see reports of 1,000 lbs. or even
more to the acre, but these are exceptional crops. It is
probable that a ton to 3 acres is a liberal estimate of the
yield for this district. Some growers claim that it can
be raised for $40 a ton, which I have no doubt is true in
exceptional cases, but it will not do to accept it as the average ; no doubt $50 to $60 the ton will come nearer the
actual cost, one year with another.

WHAT A RETIRED GROWER SAYS.

A correspondent near Homer, Ill., who was formerly
largely engaged in growing Broom-corn, writes that he
has been for some years out of the business.

" However, I will do the best I can to answer your
inquiries. I know of no one controlling more than five
or six hundred acres. I know but little of the varieties
now grown ; when I was in the business, we thought we
could grow a much better quality of brush from seed
brought from the East, than from native or acclimated
seed. I have always planted in drills. The cultivation
required is the same as for corn, except that more hand-
hoeing is necessary to keep the weeds and grass down ;
it grows extremely slowly whilst young, but rapidly after
the first foot in hight is obtained.

" One man should do all the labor necessary to the
cultivation of say 20 acres, but as the harvesting season is
short, four or five times the help would be required for
the harvesting and preparing for market. In my time we
found the average cost per acre, including rental of land,
cultivation, harvesting, and preparation for market, about

$15 per acre ; I suppose 20 to 25 per cent should be added to that for the present cost. The average yield per acre is about 600 lbs. The brush should not be bleached, but cut and cured in such a manner as to obtain the greenest color possible. It is usually cured in open buildings, spread upon poles or racks ; it is sometimes (to facilitate the curing), placed in the sun a few hours previous to being put under cover ; this, however, injures the quality. To obtain the best quality of brush or hurl, the seed should be in blossom or milk when taken off, consequently of little value except for manure. The stalks are worth something for cattle to graze upon, half as much perhaps as corn-stalks. Manure would increase the crop about as it would corn. The same land can be cultivated year after year about as corn."

MAKING BROOMS.

The manufacture of Brooms, like other mechanical trades, can be best learned by practical instruction in the needed manipulation. Still, the manufacture is a simple one, and a person of fair mechanical ingenuity can learn to make brooms from a description of the process, and by practice, become in time expert at the trade.

HOME-MADE BROOMS.

Thrifty farmers—those who look to the small economies—never buy that which can be *as well* made on the farm. By "as well" we mean not quality alone, but that the time may be profitably expended in doing the job in hand. It is very poor economy to give to making an article the time which, if expended on the regular work of the farm, would earn more than enough money to pay for the article and time expended in procuring it. In the case of brooms, it is often cheaper to raise a small patch of Broom-corn, and have the boys make it up on rainy days, than to buy the brooms ready made. While home-made brooms may not be as handsome as the "boughten" ones, they will do quite as good work—provided the right person is at the other end of the handle—as those finished off with shining wire. We recollect the remark of one of our Ohio friends, who said: "When my daughters want a new broom, they go to the shop and make one." This implies that the work is not

(45)

beyond the strength of a Buckeye girl, and that he was provident enough to keep them supplied with brush.

Those who raise only a family supply of Broom-corn, and cannot avail themselves of any of the scrapers or cleaners mentioned elsewhere, can use the very simple one shown in figure 12. It is simply a comb, made by sawing teeth in the end of a board, and nailing it firmly against a bench or other support; by drawing the brush across this the seed may be readily removed. The stalk should be cut off at about six inches from the brush; when it is ready to make up.

Fig. 12. The following directions, prepared by a friend, who lived in Maine, appeared in the *American Agriculturist* several years ago, and are founded upon his own practice.

When ready to go to work, take as much as will be needed for the number of brooms to be made, and set the stalk portions in water up to the brush, and leave it to soak an hour or two. When softened, gather in the hands enough for a broom, with the largest and best stalks on the outside, in regular order. The good appearance of the broom when finished, will depend upon the evenness of the brush and proper arrangement of the outside layers. Next, fasten a strong small cord to the ceiling, with a loop for the foot in the lower end, or tie a stick to the cord as a sort of treadle, upon which to place the foot. Wind this cord two or three times around the brush, as shown in figure 13. Grasp the brush firm-

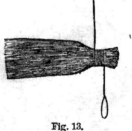

Fig. 13.

ly in both hands, and roll it round several times, increasing the pressure with the foot. Instead of the foot, some use a lever upon the lower end of the cord, one end of the lever being placed under the work-bench, and the other held by a boy, who can give the required pressure. The next operation is to wind on a strong twine for a space of $1^1/_2$ or 2 inches. This is best done by rolling the pressing cord close up next to the brush, wind the twine on, and roll off the cord towards the end, following it with the twine. To make a neat knot at the end, double one end of the twine and lay it along the outside of the stalks as shown in figure 14, letting the loose end lie out at the left. When the twine is all on, slip the right end through the loop, and draw the left end so as to bring the loop in under the coil of twine ; then cut off the two ends close in to the coil. No knot will now be visible, as the loop is out of sight, and the ends are securely fastened.

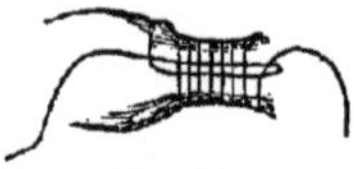

Fig. 14.

If a flat broom is to be made, which is usually the desirable form, press the brush part between two narrow boards fastened near together at one end with a piece of strong leather nailed on very securely. Figure 15 is a diagram showing the edges of the boards, as if looked down upon. The other end of the boards may be held together with a string. Instead of these boards, the brush may be put between two short boards, and screwed into a vise. The sewing is the next step. For this, a large needle of iron or steel will be required, or one of strong hard wood will answer (fig. 16), it should be six to eight inches in length. At the point where you wish to fasten the brush portion, say three or four inches below the winding cord, wind a twine once, or better twice around, and tie it firmly, leaving enough of one end to sew with. Now sew through and through the brush, letting the twine at each stitch pass around the portion you have tied on, as shown in figure

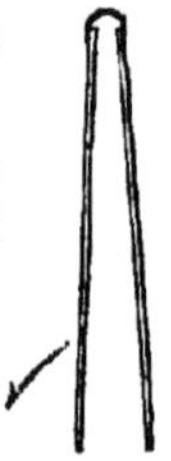

Fig. 15.

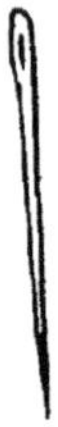

Fig. 16.

17.   Point the needle forward in making each stitch, so
as to have it come out on the opposite side a little further
along each time.   A second twine may be tied around,
and a second sewing may then be made further towards
the lower end.   Three sewings are sometimes made.
Two will generally be enough, except where the brush is
very long.   The broom is now ready for its handle.   To
put this in place, sharpen the lower end of the handle,
and drive it exactly in the center of the neck of the
broom, and fasten it with two small nails upon opposite
sides, and the broom is
complete.   The lower
ends of the brush may
need clipping a little to
make them even.   With
a little practice a very

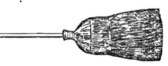

Fig. 17.—THE BROOM.

neat broom may thus be made.   They may be made still
more tasteful, though not stronger nor more durable, by
using wire instead of twine, and by paring down the
stalks, so as to make a smaller, neater shank.

ANOTHER METHOD.

A short time after the publication of the foregoing, a
communication was received from Mr. John Bennet, of
Ripley Co., Ind., giving his method of working.   He
says : " Put the but-ends of the brush in warm water to
soak awhile.   When sufficiently softened, tack one end
of a strong twine to the broom handle, about three inches
from its lower end.   Fasten the other end of the string,
which is about two feet long, to a small round stick upon
which you step with *both* feet, as shown in the engraving,
figure 18.   Lay on the brush, one stalk at a time, and
give the handle a turn sufficient to hold each new stalk
firmly.   Continue putting on and winding, until three
layers have been secured, pulling upward as the handle

is turned to tighten the string. Now commence another row nearer the lower end of the handle, and proceed as before, finishing the third course or tier with the longest and finest brush. Wind the cord around snugly a few times after the brush is all on, and fasten the end with a carpet tack. To make a broad or flat broom, more of the brush may be put upon two opposite sides than upon the other portions. Then tie the two ends of a string

Fig. 18.—WINDING THE BROOM.

the right length, slip it over the handle, and to a suitable place upon the broom, and sew it as already described in your previous article. You now have as neat a broom as you can buy, and stronger than most of those in market. With a little practice they can be made very quickly. When a boy I thought I was making money when manufacturing brooms at a dime a piece. Wire can be used instead of twine."

MAKING BROOMS BY MACHINERY.

The broom manufacture is one of those industries in which the labor may be divided with great economy. Domestic manufactures enter into competition with such industries under unfavorable conditions. It can never be hoped that the supply of brooms required by the trade, or any material portion of it, can be produced in the homes of the farmers who grow the brush. They have done their portion of the divided labor when they have provided the raw material. Yet it is true that the spare hours of the winter season, when farm labors are in good part suspended, may be in many cases profitably occupied in working up some portion of the crop. Boys and girls may give a helping hand, and earn a sum which will add a great deal to the general comfort. There is no good reason why a farmer's family might not turn out a sufficient stock of brooms to purchase most of the family groceries, or to procure a goodly supply of books and papers. Well-made brooms are usually worth at wholesale about twenty-five cents each. A pound and a half of brush will make a broom, and the handles and wire needed, cost but five or six cents. This is the whole money outlay required. The result is that an acre of brush yielding say 600 pounds, will make 400 brooms, worth $100, with an outlay for material of $24.

The following description of the manufacture, which appeared in the *American Agriculturist* for Sept., 1873, was prepared by one of the editors after a careful study of the operations carried on in one of the large broom-making establishments. The illustrations of the machinery are from sketches taken at that time. It will be seen that the machines, while they allow the brooms to be made much more rapidly and neatly than by hand, are quite simple, and can be constructed without difficulty by any one with fair mechanical ability, or the engravings here given will serve as sufficient guides for a carpenter or other mechanic to work by.

The first step is to sort the brush into three sizes, with straw of 15, 17, and 19 inches long respectively; rough, short, or crooked brush is used for the inside of the brooms, and is to be kept by itself. That which is longer than 19 inches is called "hurl," and is used for the largest and highest priced brooms, the stalk being cut off, and the straw only being used. Then the brush is cleaned

Fig. 19.—DRAINING THE BRUSH.

from any adhering seeds or hulls or broken straw, by exposing it in handfuls to a rapidly revolving drum or cylinder in the machine, used in preparing the brush for market, shown on page 28. In a small way this may be done by a coarse comb. The brush is then tied up in bundles, and the buts dipped in water and placed on a bench to drain, as shown in figure 19. The stalks are then soft and pliable, and the brush is ready for the wrapping-machine, shown at figure 20. It consists of a

table with a projecting wing at the right hand. Beneath
the top of this part of the table is a barrel or socket (*a*),
which is revolved by means of a strap from the treadle
(*b*). This treadle is made of two round pieces of board,
through the center of which an iron axle passes, and is
keyed tightly. The round boards have a number of holes
bored in them only half way through, and in these holes
round wooden rods are fitted, which serve as the steps of

Fig 20.—A BROOM-MAKING MACHINE.

the treadle, and by which it is turned by the workman's
foot. There is a pulley fastened upon the inner end of
the treadle, from which a band passes to the barrel *a*.
The broom-handle is placed in this barrel, with seven
or eight inches of the but exposed, and held fast by a set-
screw. A tack is driven part way in, about an inch and
a half from the end of the handle, and the wire wound
around it ; the tack is then driven down, and the wire
thus fastened. The handle is revolved two or three
times to give the wire a firm hold around it before any

brush is put on. The wire is wound on a reel, shown in the engraving at *c*, passes around three pulleys, by which the requisite tension is procured, and then passes to the broom-handle. When the wire is properly fastened, the operator takes a handful of coarse, rough brush, and holds the stalks beneath the wire as the handle turns, spreading them smoothly, and pounding them down closely with a flat pounder, made something like a common potato masher, which is used in kitchens, but is flat or oval instead of round. This brush is the filling, and about three small handfuls are needed for each broom. The wire should be wound around the filling three or four times, and as the brush revolves the stalks are smoothed off with a sharp knife just above the last turn of the wire. The wire is then slipped off of the brush on to the handle, and wound around it once about half an inch above the smoothed end of the stalks. Then a handful of the sorted brush, suitable for the kind of broom to be made, is taken in the left hand, and with the knife the stalks are cut half through with a sloping cut half an inch above the straw, and the half of the stalk split off. The stalks are then placed beneath the wire so that it may be wound exactly over where they were cut. The treadle is turned until the stalks are all bound on, when another handful is taken and treated precisely the same way, and then finally another handful. Each handful consists of six or eight stalks, and they should be placed smoothly and close together under the wire. The wire is bound evenly around the stalks until there is sufficient to hold the broom firmly together, when it is fastened with a tack as at the commencement. The pounder is constantly used to pack the brush beneath the wire and make the broom firm and hard.

It would be a good practical lesson to take an old broom to pieces while studying these processes, so as to fix them clearly on the mind. The broom is now of a

round shape, and needs to be made flat and sewed. This
is done in the clamps shown at figure 21. These are
simply a pair of wooden jaws, very similar to those used
by harness-makers in which the leather is held to be
sewed. The broom is put into the clamps, which are
pressed together by the lever which is shown projecting

Fig. 21.—SEWING THE BROOMS.

at the side (d). The lever is made of a piece of wood
with a handle, and an iron band is fastened to it so as to
embrace the jaws of the clamp. The band is pivoted to
one of the clamps, and when the handle is lifted, the
other clamp is forced forward towards the first one, and
pinches whatever may be placed between them. The
movable clamp is hinged to the floor. The clamps are
about eight inches broad. Before being squeezed in the

clamps, the brush is arranged and put into proper shape.
Then there are two or three guides, made of iron, with
curved jaws, hinged on to each side of the clamp shown
at *e, e,* in the figure. When these are turned up against
the broom, they show the exact place where the sewing
should be done ; when the broom is sewn only in two
places, but two guides are used, when sewn three times,
three are needed. A supply of twine cut to the proper
length is on the table ; a length of it is taken and passed

Fig. 22.—TRIMMING THE BROOMS.

by means of a long needle through the broom, from the
left-hand side, about an inch or less from the edge. The
knot by which the end of the twine is tied is drawn just
inside of the brush, so that it is not seen ; then the
twine is passed twice around the broom and drawn tight,
the guide keeping it in its proper position. The needle
is then passed through and through the broom, under
and over the twine each time, making stitches about an

inch apart, until they cross the broom. Then another guide
is turned up which reaches about an inch nearer the bottom
of the broom, and another double turn of twine is made,
and more stitches, and this is repeated in long-straw
brooms yet once more. It is only necessary then to trim
the broom smoothly, which, where large quantities are
made, is done by the machine shown at figure 22 ; but
in other cases may be done by means of a sharp knife or
a pair of sheep-shears, to finish it ready for market. The
wire used on the best brooms is known in the trade as
"tinned broom wire"; cheaper brooms are wound with
"bright annealed iron wire," the sizes Nos. 29 to 36 being used.

Broom handles sell at from $16 to $18 per thousand,
and are retailed in small lots at 2 to 3 cents each. In
cities they are kept by the dealers in wooden ware.

Brooms are packed in bunches of one dozen each, being
sewed together through the brush, and bound by a cord
at the ends of the handles, and the manufacture is com-
pleted.

The value of the brooms in the market of course depends
altogether upon the material of which they are made.
Those made of poor yellow brush, which is so short that
the stubs of the brush are used in making the brooms,
bring but half the price of those made of the long green
hurl. This should be remembered when the brush is
harvested, as well as when the choice of seed is made for
planting. The average prices of brooms, as quoted in
May, 1876, were $2 to $3.50 per dozen.

EXTENT OF THE MANUFACTURE.—Very few have any
idea of the extent of this apparently unimportant manufacture. There are in the United States 625 factories
engaged in making brooms and whisks, employing 5,206
hands. The amount of capital invested in the business,
is estimated at a little over $2,000,000. The annual

amount paid for brush. is $3,672,837. The value of brooms of all kinds manufactured, is $6,622,285.

But very little brush is exported, an insignificant amount being sent to Cuba and South America, but the export of ready made brooms is annually increasing. The value of the brooms exported in 1873, was $131,319; this increased in 1874 to $170,185, and in 1875 it reached to $204,696.

Gardening for Pleasure.

A GUIDE TO THE AMATEUR IN THE

FRUIT, VEGETABLE, AND FLOWER GARDEN,

WITH FULL DIRECTIONS FOR THE

GREENHOUSE, CONSERVATORY, AND WINDOW-GARDEN.

BY PETER HENDERSON.

AUTHOR OF "GARDENING FOR PROFIT," AND "PRACTICAL FLORICULTURE."

Illustrated.

EDITORIAL NOTICES.

ONE of the most popular works of recent years on similar topics was the "Gardening for Profit" of Mr. PETER HENDERSON, the well-known florist of Jersey City. He has been equally fortunate in the title of a new book from his pen, just published by the ORANGE JUDD CO., of New-York —"Gardening for Pleasure." The author has a happy faculty of writing for the most part just what people want to know—so that, although his books are neither exhaustive nor especially elaborate, they proceed to the gist of the subject in hand with so much directness and simplicity that they fill a most important and useful sphere in our rural literature.—*The Cultivator and Country Gentleman, Albany, N. Y.*

IT gives, in a clear, intelligible form, just the information that novices and even experienced cultivators wish to have always accessible, and will be specially valuable to those who keep house plants.—*The Observer, New-York City.*

MR. PETER HENDERSON has followed up "Gardening for Profit" with "Gardening for Pleasure," into which is packed much useful information about window-gardens, the management of flower-beds, etc.—*The Independent, New-York City.*

HE is a thoroughly practical man, uses plain, common language, and not technical terms, in his statements and explanations, and puts the staff of knowledge directly into the hands of the amateur and sets him at work. —*The Press, Providence, R. I.*

PEOPLE who have money to spend in adorning their grounds, are told here how to do it to the best advantage, and ladies are fully instructed in all the art and mystery of window-gardening. It will prove a useful guide to all who have a taste for flowers, and also contains practical instructions for the cultivation of fruits and vegetables.—*The Transcript, Portland, Me.*

THIS volume is eminently clear in its style and practical in its directions. Its appearance is timely, as it contains some valuable hints upon winter flowering plants and their proper cultivation, together with plain directions how to raise them from seed and to multiply them by cuttings.— *Courier-Journal, Louisville, Ky.*

Price, post-paid, $1.50.

ORANGE JUDD COMPANY,

245 Broadway, New-York.

PRACTICAL FLORICULTURE,

A GUIDE TO THE
SUCCESSFUL PROPAGATION AND CULTIVATION OF
Florists' Plants.
BY PETER HENDERSON, BERGEN CITY, N. J.

Mr. HENDERSON is known as the largest Commercial Florist in the country. In the present work he gives a full account of his modes of propagation and cultivation. It is adapted to the wants of the amateur as well as the professional grower.

The scope of the work may be judged from the following

TABLE OF CONTENTS.

ILLUSTRATED.
SENT POST-PAID. PRICE, $1.50

ORANGE JUDD & COMPANY,
245 BROADWAY, New-York

THE SHEPHERD'S MANUAL.

A Practical Treatise on the Sheep.

DESIGNED ESPECIALLY FOR

AMERICAN SHEPHERDS.

By HENRY STEWART.

Illustrated.

THIS Manual is designed to be a hand-book for American shepherds and farmers. It is intended to be so plain that a farmer, or a farmer's son, who has never kept a sheep, may learn from its pages how to manage a flock successfully, and to be so complete that even the experienced shepherd may gather some suggestions from it. The results of personal experiences of some years with the characters of the various modern breeds of sheep, and the sheep-raising capabilities of many portions of our extensive territory and that of Canada, most of which have been visited with a view to the effects upon our sheep of the varying climate and different soils; and the careful study of the diseases to which our sheep are chiefly subject, with those by which they may eventually be afflicted through unforeseen accidents; as well as the methods of management called for under our circumstances, were finally gathered into the shape in which they are here presented to the shepherds of America, with the hope that they may be as acceptable and useful to them as they would have been, when he first undertook the care of a flock, to THE AUTHOR.

CONTENTS.

Price, post-paid, $1.50.

ORANGE JUDD COMPANY,

245 Broadway, New-York.

NEW AMERICAN FARM BOOK.

ORIGINALLY BY

R. L. ALLEN,

AUTHOR OF "DISEASES OF DOMESTIC ANIMALS," AND FORMERLY EDITOR OF
THE "AMERICAN AGRICULTURIST"

REVISED AND ENLARGED BY

LEWIS F. ALLEN,

AUTHOR OF "AMERICAN CATTLE," EDITOR OF THE "AMERICAN SHORT-HORN
HERD BOOK," ETC.

CONTENTS:

SENT POST-PAID, PRICE $2.50.

ORANGE JUDD & CO.,

245 Broadway, New-York

Ingram Content Group UK Ltd.
Milton Keynes UK
UKHW022214110723
424974UK00004B/37